FARM MACHINES

BALER

KATHY MIDDLETON

A Crabtree Roots Book

Crabtree Publishing
crabtreebooks.com

School-to-Home Support for Caregivers and Teachers

This book helps children grow by letting them practice reading. Here are a few guiding questions to help readers build comprehension skills. Possible answers appear here in red.

Before Reading:

- What do I think this book is about?
    - *I think this book is about balers.*
    - *I think this book will talk about what balers do.*
- What do I want to learn about this topic?
    - *I want to learn about what jobs balers do.*
    - *I want to learn why balers are important farm machines.*

During Reading:

- I wonder why...
    - *I wonder why farmers use a baler.*
    - *I wonder why balers are needed on a farm.*
- What have I learned so far?
    - *I have learned that balers attach to the back of a tractor.*
    - *I have learned that balers have a bale chamber.*

After Reading:

- What details did I learn about this topic?
    - *I have learned that balers gather and bind hay and straw.*
    - *I learned that bales can be made round or in rectangles.*
- Read the book again and look for the vocabulary words.
    - *I see the word **baler** on page 3 and the word **tractors** on page 7. The other vocabulary words are found on page 14.*

This is a **baler**.

Balers gather and bind straw and hay into **bales**.

Balers attach to the back of **tractors**.

Bales are
made inside a
bale chamber.

Bales can be made round or in **rectangles**.

Bales keep hay
and straw fresh for
animals on the farm.

Word List

Sight Words

a	bind	inside	or
and	can	into	round
animals	for	is	straw
are	fresh	keep	the
attach	gather	made	this
back	hay	of	to
be	in	on	

Words to Know

bale chamber

baler

bales

rectangles

tractors

46 Words

This is a **baler**.

Balers gather and bind straw and hay into **bales**.

Balers attach to the back of **tractors**.

Bales are made inside a **bale chamber**.

Bales can be made round or in **rectangles**.

Bales keep hay and straw fresh for animals on the farm.

FARM MACHINES BALER

Written by: Kathy Middleton
Designed by: Under the Oaks Media
Proofreader: Crystal Sikkens
Production manager: Candice Campbell
Prepress technician: Katherine Kantor

Photographs: All images from Sutterstock

Crabtree Publishing

crabtreebooks.com 800-387-7650
 In Canada: We acknowledge the financial support of the Government of Canada through the Canada Book Fund for our publishing activities.

Printed in the U.S.A./CP2025

Published in Canada
Crabtree Publishing
616 Welland Ave.
St. Catharines, Ontario
L2M 5V6

Published in the United States
Crabtree Publishing
347 Fifth Ave
Suite 1402-145
New York, NY 10016

Library and Archives Canada Cataloguing in Publication
Available at Library and Archives Canada

Library of Congress Cataloging-in-Publication Data
Available at the Library of Congress

Hardcover: 978-1-0398-8102-0
Paperback: 978-1-0398-8463-2
Ebook (pdf): 978-1-0398-8222-5
Epub: 978-1-0398-8342-0
Audio: 978-1-0398-8505-9
Read-Along: 978-1-0398-8547-9

Word Count: 46
GRL: D
Subject: Social studies
Genre: Nonfiction
Standard: D2.Civ.2.K-2.
Vocabulary Words: baler, bale chamber, bales, rectangles, tractors

BALER

Early readers will be introduced to balers through simple text and vibrant images to aid comprehension. Learn all about the characteristics of farm machines.

Books in ***Farm Machines*** include:

Baler
Combine
Crop Sprayer
Cultivator
Planter
Plow
Swather
Tractor

Crabtree Roots

These books feature simple text with each picture to aid comprehension and help children learn to read with confidence. Every book begins with a list of questions to help caregivers and teachers guide young readers.

Teacher's Guides

Crabtree Publishing
crabtreebooks.com

ISBN 978-1-0398-8463-2
90000 >
9 781039 884632

FARM MACHINES

CULTIVATOR

KATHY MIDDLETON

Crabtree Roots